APPLIED RESEARCH FOR CORN PRODUCTION IN INDIANA, 2022

APPLIED RESEARCH FOR CORN PRODUCTION IN INDIANA, 2022

DANIEL QUINN

PURDUE UNIVERSITY PRESS
WEST LAFAYETTE, INDIANA

Cataloging-in-publication data on file at the Library of Congress.

978-1-62671-325-3 (paperback)
978-1-62671-326-0 (epdf)

Cover image: Purdue University Department of Agriculture

CONTENTS

ACKNOWLEDGMENTS

This report entails a detailed summary of applied field research trials for corn production systems in Indiana, conducted under the direction of Dr. Daniel Quinn and the Purdue Corn Agronomy team in the Department of Agronomy at Purdue University. The authors extend many thanks to the Purdue Agronomy Center for Research and Education, the Purdue Agricultural Centers, farmer cooperators, and the many industry collaborators and funding agencies who help provide the necessary resources needed to support this research. Special recongnition is extended to Ana Morales, Riley Seavers, Malena Bartaburu, and Nathaly Vargas who assisted with trial organization, data collection and processing, and the preparation of this report. In addition, the authors also extend thanks to Crystal Paris for report booklet design and visiting scholars and undergraduate students Erick Oliva, Narciso Zapata, Lizeng Zhao, Caroline Carlton, and Sergio Rubiano who assisted with trial organization, data collection, and scouting. Overall, the combined efforts of various colleagues, professionals, students, and farmers are responsible for the success of this research.

The authors would also like to thank those below for their support in 2022:

Indiana Corn Marketing Council	USDA-NIFA
Corteva Agriscience	John Deere
Pioneer	Netafim
Bayer Crop Science	Copperhead Ag
FMC	Purdue University
Ceres Solutions	Winfield United
NRCS CIG	Becks
BASF	

SUMMARY OF THE 2022 CORN GROWING SEASON IN INDIANA

In 2022, Indiana produced a statewide corn yield average of 191 bushels per acre (bu/ac), which marks the second highest corn yield average produced in the state behind 2021 (195 bu/ac) (Figure 1). Crop condition rated good to excellent averaged 56% throughout the entire growing season, with a large decrease observed from the beginning of June (76% good to excellent) to the beginning of July (48% good to excellent; USDA-NASS, 2022). The 2022 season was highlighted by a delayed start due to wet conditions at the start of planting (11% planted as of May 9, 2022, 28 percentage points behind the 5-year average) and an immediate switch to dry conditions in June, which dramatically reduced crop conditions ratings. For example, the Purdue University research farms in West Lafayette, IN and Lafayette, IN received only 1.2 inches and 0.6 inches of precipitation, respectively, in June. Despite stressful conditions observed across the state in June, disease pressure remained low and minimal pollination issues were observed due to timely July rainfall. In addition, moderate temperatures and adequate rainfall in both August and September likely improved grain fill conditions, causing above average kernel weight numbers and leading to higher-than-expected yields across Indiana.

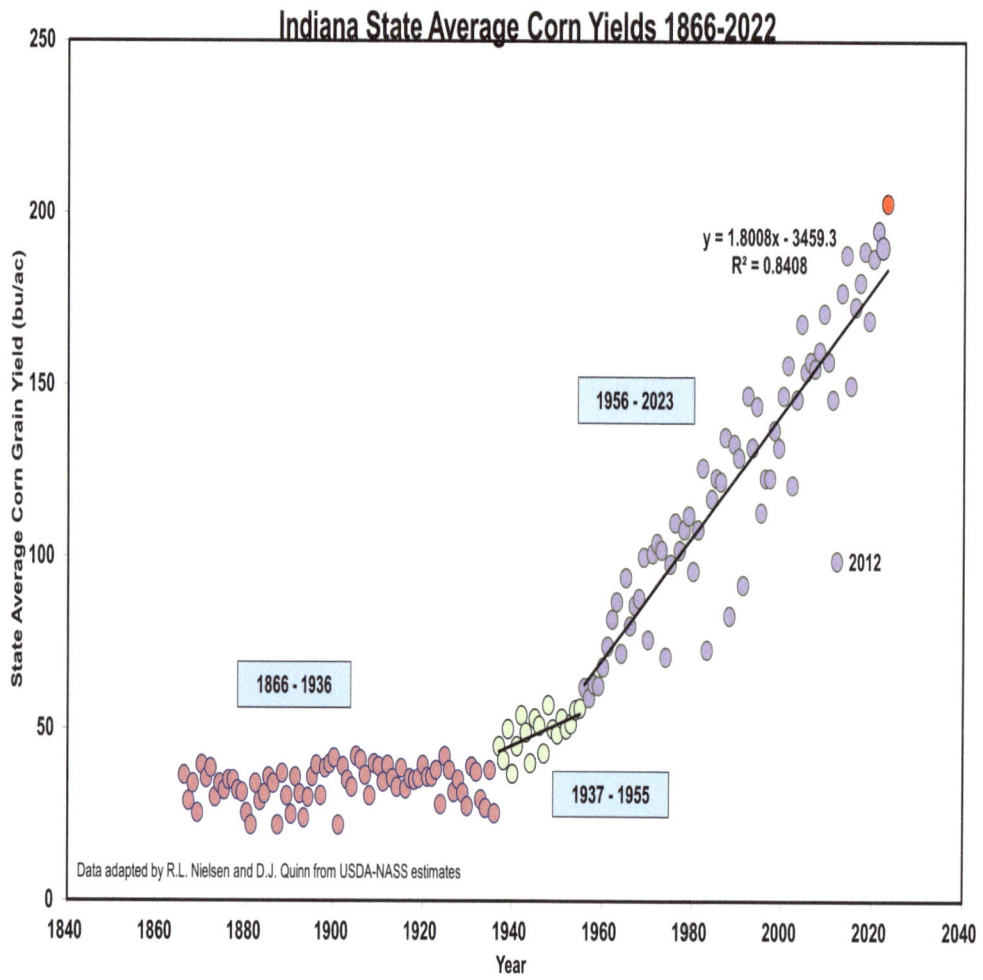

FIGURE 1. Historic state average corn grain yield trends for Indiana (1866–2022).

RANDOMIZATION, REPLICATION, AND STATISTICS

MAKING SENSE OF APPLIED FIELD RESEARCH

Field research trials are an important part of understanding how specific agronomic practices can improve farm productivity. Universities such as Purdue use both research station and on-farm research trials across the state to help drive our recommendations and provide management information for Indiana farmers. However, some of our research practices and conclusions may differ from various private-sector research trials and potentially what you may see on your own farm. For example, you may ask: "Why did they set up the research trial that way?," "What are those letters next to the yield values they are presenting?," and "Why does it seem the university never sees any yield responses from various products?" Therefore, it is important to understand how we approach field research trials, the steps we take to determine our conclusions, and how understanding these approaches can help you better understand and test practices more accurately on your own farm.

Two of the first questions I often ask people when discussing research results are: (1) Do you have a yield monitor in your combine? and (2) When traveling across the field during harvest, do those yield values stay the same? The answer I receive 100% of the time is no (if yes, you may need to consider a new monitor), and this is largely due to the variability throughout the field caused by soil type differences, elevation differences, and so on. Therefore, when setting up field research trials we often designate a treatment (e.g., new product) and compare that to a nontreated control (e.g., business as usual). And two of the most important questions we ask after harvest are: (1) Was the yield difference observed truly caused by the product we applied? (2) Or was the yield difference only due to the treated areas being in a more productive part of the field? For example, in Figure 2, if I split a field in half and apply my treatment on one half of the field but don't apply my treatment on the other half of the field, I may find a yield difference of 15 bushels per acre and think to myself, "I should apply this product on all of my acres." However, when you look closer, it is easy to see that the treated area of the field encompassed a larger portion of one soil type, whereas the nontreated area encompassed a larger portion of another soil type. Therefore, it is difficult to determine whether the yield response was due to the product applied or due to the treated area being in a more productive area of the field.

In our university research trials, we approach testing a treatment within a field using randomization, replication (repetition of an experiment in similar conditions), and statistics (Figure 2 and Table 1). For example, compare Figure 2 and Figure 3. Figure 2 highlights how we typically set up one of our research trials using replication and randomization of the treated and nontreated passes to account for field differences. Each of these practices helps us improve the reliability of our conclusions, account for random error (e.g., field variability), and determine the true causes of yield differences observed. Furthermore, it is also important for us to perform these research trials across multiple locations and multiple years to determine how treatment responses

FIGURE 2. Example of a split-field comparison between a nontreated control and a designated treatment.

FIGURE 3. Example of a replicated and randomized field research trial comparison between a nontreated control and a designated treatment.

TABLE 1. *Corn grain yield comparisons between the nontreated control and an imposed treatment following a randomized and replicated field research trial.*

TREATMENT	YIELD (BU/AC)
Nontreated	204 a*
Treated	208 a

* Average yield values that contain the same corresponding letters are not statistically different ($P > 0.1$) from each other.

may differ in different fields and different environments. We also use statistical models to help determine our conclusions (Table 1). Using statistics helps us determine if the differences we detect are due to random error, or due to the treatment we tested. For example, if you have ever seen university data presented (or the data presented in this report), you have probably seen data presented similar to Table 1. At first glance, after we randomized and replicated our treatments (Figure 3), the treated areas seem to have increased corn yield by 4 bushels per acre (Table 1). However, our conclusions suggested no yield differences were observed. Therefore, through the research steps we implemented, it was determined that the yield difference was due to random error (e.g., field variability) and not due to the product or management practice tested. The letters next to the yield values help us highlight where statistical (yield differences due to treatments) differences were observed.

In conclusion, when testing a new product or practice on your own farm, it is important to think about how to design and set up a trial to accurately test the new product or practice. Just because you observe a yield difference doesn't mean the new product or practice you tested is the reason for this difference. At Purdue, it is our goal to accurately assess new products and practices to determine whether or not these are truly the reason behind observed yield differences. In addition, as you sit in on various meetings and presentations and examine research results, ask yourself: How did they design and set up this research trial? Did they use randomization, replication, and statistics, and if not, are the yield differences being discussed truly due to the product applied? Over how many different environments and years was this product tested? Understanding and asking these questions can help you determine the best products and management practices to implement and improve your operation.

AGRONOMY CENTER FOR RESEARCH AND EDUCATION (ACRE)

CORN RESPONSE TO INPUT-INTENSIVE MANAGEMENT (ACRE)

Daniel Quinn: Department of Agronomy, Purdue University
Malena Bartaburu: Department of Agronomy, Purdue University
Darcy Telenko: Department of Botany and Plant Pathology, Purdue University
Steven Brand: Department of Botany and Plant Pathology, Purdue University
Rachel Stevens: Purdue Agronomy Center for Research and Education (ACRE)

Study Location: West Lafayette, IN
Soil Type: Chalmers silty clay loam (0–2% slope)
Planting Date: May 14, 2022 | **Harvest Date:** October 19, 2022
Corn Hybrid: Pioneer P1185Q | **Corn Seeding Rate:** 30,000 and 36,000 seeds/ac
Corn Nitrogen (N) Fertilizer Rate and Source: 180 lbs N/ac, UAN (28-0-0)
Previous Crop: Soybean | **Tillage:** Conventional
Study Replications: 5

RESEARCH TRIAL OVERVIEW:

A field research trial was established at the Purdue Agronomy Center for Research and Education (ACRE) in Tippecanoe County, IN. The research trial examined corn yield response to different management practices and inputs and was utilized to analyze the impact of each one applied individually and in combination. The trial was designed as a randomized complete block design with eight treatments and five replications. Plots measured 10 feet wide (four 30-inch corn rows) by 30 feet long, and the center two rows were harvested with a small-plot combine and adjusted to 15.5% moisture for yield analysis. All plots received a pre-plant N application totaling 40 lbs N/ac.

RESEARCH TREATMENTS:

1. Control treatment (C) based on Purdue University seed rate (30K seeds/ac) and nitrogen (N) fertilizer rate recommendations.
2. C + surface banded starter (2x0) fungicide (flutriafol, Xyway LFR, 15 oz/ac)
3. C + 20% increase in corn seeding rate (36K seeds/ac)
4. C + sulfur fertilizer [5.2 gallons/ac as ammonium thiosulfate (ATS) at V5 sidedress]
5. C + foliar micronutrients (zinc, manganese, and boron applied at the V6 growth stage)
6. C + late-season N application [starter N (2x2) + V5 sidedress N (60% remaining N rate) + V10-12 growth stage sidedress N surface-banded with drop tubes on a sprayer (40% remaining N rate), total N rate remained the same as other treatments]
7. C + foliar fungicide applied at the R1 growth stage (prothioconazole, trifloxystrobin, fluopyram, Delaro Complete, 10 oz/ac)
8. Intensive treatment: All additional inputs and management practices applied together

RESULTS:

TABLE 2. *Corn grain yield, grain moisture, treatment cost, and net profit differences observed from applied treatments in 2022. West Lafayette, IN.*

TREATMENT	GRAIN MOISTURE	GRAIN YIELD	TREATMENT COST[†]	NET PROFIT[†]
	--- % ---	-- bu/ac --	-- $/ac --	-- $/ac --
Control	17.8 b*	206.1 c	253.4	1119.34 bc
C + 2x0 Fungicide	17.4 b	206.3 c	274.3	1092.9 c
C + Increased Seed	17.4 b	213.2 bc	272.7	1141.9 bc
C + Sulfur	17.3 b	210.1 c	270.7	1128.2 bc
C + Foliar Micro	17.6 b	213.9 bc	274.1	1150.2 abc
C + V10-12 SD N	18.0 b	232.4 ab	261.1	1286.7 a
C + R1 Fungicide	18.0 b	226.5 ab	277.9	1230.8 a
Intensive	19.1 a	243.3 a	356.1	1264.5 a
P-value	*0.092*	*0.043*	—	*0.023*

* Mean values that do not contain the same corresponding letter are determined statistically different (*P* < 0.1).

† Treatment costs were calculated as the combined cost of corn seed, fertilizer cost, chemical input cost, and application cost. Prices were calculated as an average from various local retailers. Net profit was calculated based on average harvest corn grain cash price ($6.66) + average grain yield – treatment costs.

SUMMARY (TAKE-HOME POINTS):

• The R1 foliar fungicide application, late-season sidedress N application, and intensive treatment (all inputs applied) significantly increased corn grain yield and net profit in comparison to the control in 2022 at this location (Table 1).

• Foliar disease presence (e.g., tar spot, gray leaf spot, northern corn leaf blight) was significantly reduced by the R1 fungicide application, likely driving the observed yield responses (Table 1).

- Late-season N application responses were likely due to increased plant N uptake and N use efficiency due to more timely rainfall following the V10–12 N application as compared to observed June drought conditions, which drove visual N deficiency symptoms and limited plant uptake following the V5 sidedress N only treatments.
- No significant yield responses were observed from the 2x0 starter fungicide, 20% increase in corn seeding rate, and sulfur application at this location in 2022.

CORN RESPONSE TO NITROGEN FERTILIZER APPLICATION TIMING FOLLOWING A RYE COVER CROP (ACRE)

Daniel Quinn: Department of Agronomy, Purdue University

Riley Seavers: Department of Agronomy, Purdue University

Shalamar Armstrong: Department of Agronomy, Purdue University

Rachel Stevens: Purdue Agronomy Center for Research and Education (ACRE)

Study Location: West Lafayette, IN

Soil Type: Drummer Fine-Silty (0–2% slope), Raub-Brenton complex (0-1% slope)

Planting Date: May 13, 2022 | **Harvest Date:** October 10, 2022

Corn Hybrid: Pioneer P1185Q | **Corn Seeding Rate:** 30,000 seeds/ac

Corn Nitrogen (N) Fertilizer Rate and Source: 200 lbs N per acre, UAN (28-0-0) and Anhydrous Ammonia (82-0-0)

Previous Crop: Soybean | **Tillage:** No-Till

Study Replications: 4

RESEARCH TRIAL OVERVIEW:

This research trial was established at the Purdue Agronomy Center for Research and Education in Tippecanoe County, IN. This research trial assessed corn yield differences when utilizing different N fertilizer application timings following a fall-planted cereal rye cover crop and no rye cover crop. The rye cover crop was fall drill-seeded at a rate of 45 lbs/ac and was chemically terminated with glyphosate two weeks before corn planting. All plots received a fall anhydrous ammonia application at 60 lbs N/ac. Plots measured 30 feet wide (12, 30-inch corn rows) x 500+ feet long. The center eight rows were harvested with a commercial combine containing a calibrated yield monitor and were adjusted to 15.5% moisture for yield analysis.

RESEARCH TREATMENTS:

1. No cover crop + V5 sidedress N (coulter-inject, UAN 28-0-0)
2. No cover crop + V10 sidedress N (surface-banded, UAN 28-0-0)
3. No cover crop + V5 + V10 sidedress N (60% remaining N at V5 and 40% remaining N at V10)
4. Cereal rye cover crop + V5 sidedress N (coulter-inject, UAN 28-0-0)
5. Cereal rye cover crop + V10 sidedress N (surface-banded, UAN 28-0-0)
6. Cereal rye cover crop + V5 + V10 sidedress N (60% remaining N at V5 and 40% remaining N at V10)

RESULTS:

TABLE 3. *Mean cereal rye cover crop aboveground biomass and nutrient uptake values. Rye cover crop biomass was sampled immediately prior to spring termination. West Lafayette, IN.*

BIOMASS	CARBON	NITROGEN	C:N[*]
-- lbs/ac --	-- lbs/ac --	-- lbs/ac --	-- lbs/ac --
1017	420	37.7	11:1

* Carbon (C) to nitrogen (N) ratio of aboveground biomass at termination.

TABLE 4. *Mean plant population, grain moisture, and grain yield differences observed across cereal rye cover crop presence and nitrogen (N) fertilizer application timing. West Lafayette, IN.*

COVER CROP	N TIMING	V5 PLANT POPULATION	GRAIN MOISTURE	GRAIN YIELD
		-- plants/ac --	-- % --	-- bu/ac --
No Cover	0N	29,185 a*	17.3 d	197.7 d
	V5	29,185 a	18.7 c	238.6 a
	V10	29,620 a	18.6 c	229.5 b
	V5+V10	29,185 a	18.7 c	234.0 ab
Rye Cover Crop	0N	29,010 a	18.6 c	133.3 e
	V5	29,010 a	20.3 b	213.7 c
	V10	29,446 a	20.7 a	200.2 d
	V5+V10	29,446 a	20.3 b	210.7 c
P-value		0.816	0.001	0.001

* Mean values that do not contain the same corresponding letter are determined statistically different ($P < 0.1$).

SUMMARY (TAKE-HOME POINTS):

- Rye cover crop biomass at termination averaged 1017 lbs/ac, which contained 420 lbs/ac of carbon, 37.7 lbs/ac of N, and a C:N ratio of 11:1 (low C:N ratio due to fall anhydrous application) (Table 3).
- Across all treatments examined, a rye cover crop reduced corn yield by 35 bu/ac. Yield reductions were likely due to observed N stress and delayed plant growth.
- The V5 N application timing at ACRE had the highest yield value across both cover crop treatments (238.6 bu/ac in the no rye cover crop treatments and 213.7 bu/ac in the rye cover crop treatments) (Table 4).
- Grain moisture at harvest was increased by 1.6 points with rye cover crop presence (Table 4).
- Preliminary results suggest a full-rate sidedress N application at the V5 growth stage is the most beneficial when following a rye cover crop. However, yield reductions following a rye cover crop are still present and may have been exacerbated by the lack of starter N fertilizer (2x2) at planting at this location.

CORN RESPONSE TO INTENSIVE MANAGEMENT WITH IRRIGATION AND FERTIGATION (ACRE)

Daniel Quinn: Department of Agronomy, Purdue University

Nathaly Vargas: Department of Agronomy, Purdue University

Laura Bowling: Department of Agronomy, Purdue University

Katy Mazer: Department of Agronomy, Purdue University

Shaun Casteel: Department of Agronomy, Purdue University

Rachel Stevens: Purdue Agronomy Center for Research and Education (ACRE)

Study Location: West Lafayette, IN

Soil Type: Chalmers silty clay loam (0–2% slope), Toronto-Millbrook complex (0–2% slope)

Planting Date: May 31, 2022 | **Harvest Date:** October 28, 2022

Corn Hybrid: Pioneer P1185Q | **Corn Seeding Rate:** 30,000 and 36,000 seeds/ac

Corn Nitrogen (N) Fertilizer Rate and Source: 180 lbs N/ac, UAN (28-0-0)

Previous Crop: Soybean | **Tillage:** Conventional

Study Replications: 6

RESEARCH TRIAL OVERVIEW:

This research trial was established at the Agronomy Center for Research and Education (ACRE) in Tippecanoe County, IN. This research trial examined corn grain yield differences under conventional and intensive management with and without the use of recycled drainage water for application of irrigation and fertigation using surface-drip lines. The experimental design of this trial was a randomized complete block design with six replications. All plots measured 60 feet wide (24, 30-inch corn rows) x 75 feet long. All plots received a 2x2 starter fertilizer application of nitrogen at 40 lbs N/ac. The center four rows of each plot were harvested with a small-plot combine and adjusted to 15.5% moisture for yield analysis.

RESEARCH TREATMENTS:

1. Control treatment (C) based on Purdue University seed rate (30K seeds/ac) and nitrogen (N) fertilizer rate recommendations.
2. C + Irrigation. Irrigation water was applied through a surface-drip application. Irrigation applications were decided daily based on soil moisture levels.
3. Intensive Management (IM). Seeding rate of 36,000 seeds/ac, multiple in-season N fertilizer application [starter N (2x2) + V5 sidedress N (60% remaining N rate) + V10-12 growth stage sidedress N surface-banded with drop tubes on a sprayer (40% remaining N rate), total N rate remained the same as other treatments], sulfur fertilizer [5.2 gallons/ac as ammonium thiosulfate (ATS) at V5 sidedress], and foliar fungicide at the R1 growth stage (mefentrifluconazole, pyraclostrobin, Veltyma, 10 oz/ac).
4. Intensive Management (IM) + Irrigation. Treatment contained all intensive applications above combined with irrigation presence.

5. Intensive Management (IM) + Fertigation. Fertigation consisted of 20 lbs N/ac applied as UAN (28-0-0) and an additional 2 lbs S/ac applied as ATS (12-0-0-26S) injected through the surface drip lines with irrigation application at the V12, R1, and R3 growth stages.

RESULTS:

TABLE 5. *Corn grain moisture and grain yield differences observed from applied treatments in 2022. West Lafayette, IN.*

TREATMENT	GRAIN MOISTURE	GRAIN YIELD
	--- % ---	-- bu/ac --
Control (C)	19.6 a*	168.5 b
C + Irrigation	19.0 a	181.5 a
Intensive (IM)	19.3 a	183.4 a
IM + Irrigation	19.3 a	188.5 a
IM + Fertigation	19.2 a	189.9 a
P–value	0.535	0.022

* Mean values that do not contain the same corresponding letter are determined statistically different ($P < 0.1$).

SUMMARY (TAKE-HOME POINTS):

- Significant soil crusting caused by heavy rainfall immediately following planting reduced overall plant stands and caused lower than expected grain yields for this trial.
- Irrigation, intensive management, intensive management + irrigation, and intensive management + fertigation all increased corn yield beyond the control (Table 5).
- Overall, the data suggests the inclusion of irrigation water resulted in the greatest influence in corn yield beyond additional management factors (e.g., higher seed rate, sulfur, fungicide, etc.).
- Preliminary results suggest recycled drainage water applied through surface-drip lines has the potential to increase corn yield by supplementing crop water needs during the season. However, results will be repeated in the upcoming years to capture different environmental conditions and will also include a transition to subsurface drip line for further evaluation.

PINNEY PURDUE AGRICULTURAL CENTER (PPAC)

CORN RESPONSE TO VARIOUS SEEDING RATES AT PLANTING (PPAC)

Daniel Quinn: Department of Agronomy, Purdue University
Alex Leman: Pinney Purdue Agricultural Center
Gary Tragesser: Pinney Purdue Agricultural Center

Study Location: Pinney Purdue Agricultural Center, Wanatah, IN
Soil Type: Sebewa loam
Planting Date: May 24, 2022 | **Harvest Date:** Oct. 29, 2022
Corn Hybrid(s): Pioneer P1099Q
Corn Nitrogen (N) Fertilizer Rate and Source: 212 lbs N per acre, UAN (28-0-0)
Previous Crop: Soybean | **Tillage:** Conventional
Study Replications: 5

RESEARCH TRIAL OVERVIEW:

A field research trial was established at the Pinney Purdue Agricultural Center (PPAC) in Porter County, IN. The research trial examined corn yield response to different seeding rates and was utilized to add additional data toward corn seeding rate recommendations in Indiana. The trial was designed as a randomized complete block design with five replications. Plots measured 30 feet wide (12, 30-inch corn rows) by 500 feet long, and the center six rows were harvested with a commercial combine containing a calibrated yield monitor and adjusted to 15.5% moisture for yield analysis.

RESULTS:

TABLE 6. *Corn grain yield and grain moisture differences observed from the different seeding rates applied in this trial in 2022.*

CORN SEEDING RATE	GRAIN MOISTURE	GRAIN YIELD
seeds/acre	%	bu/ac
20,000	19.3 a*	207.3 c
26,000	18.9 b	233.1 ab
32,000	18.1 c	238.0 a
38,000	17.6 d	231.8 b
44,000	17.2 e	230.2 b
P-value	0.001	0.001

* Mean values that do not contain the same corresponding letter are determined statistically different ($P < 0.1$).

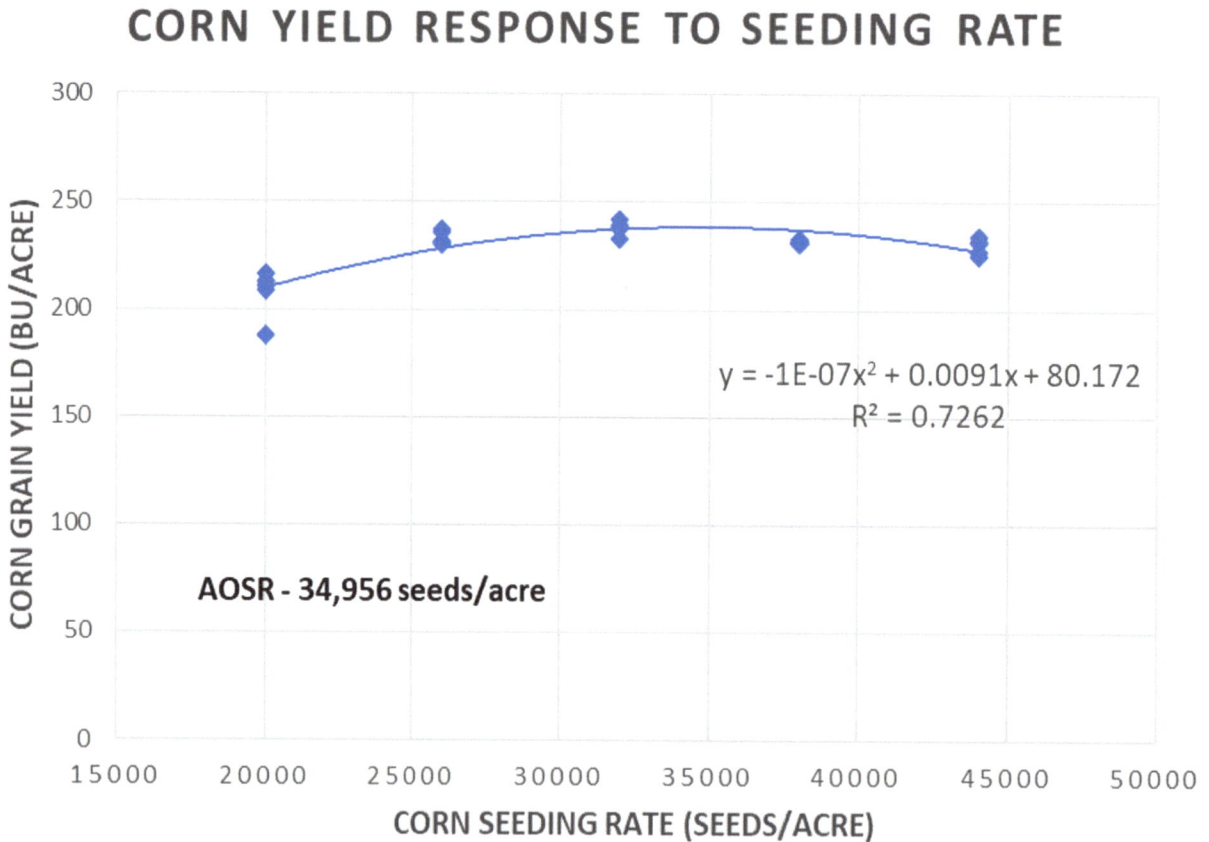

CORN YIELD RESPONSE TO SEEDING RATE

$$y = -1E\text{-}07x^2 + 0.0091x + 80.172$$
$$R^2 = 0.7262$$

AOSR - 34,956 seeds/acre

FIGURE 4. Quadratic regression analysis examining corn grain yield response to different seeding rates at planting. Agronomic optimum seeding rate (AOSR) determines the seeding rate at which grain yield was maximized using the quadratic regression equation.

SUMMARY (TAKE-HOME POINTS):

- Using quadratic regression, maximum corn grain yield was observed at a corn seeding rate of 34,956 seeds per acre at this location in 2022 (Figure 4). However, minimal yield differences were observed between 26,000 and 38,000 seeds per acre (Table 6).
- Corn grain moisture at harvest was reduced as seeding rate was increased (Table 6).
- For more information regarding corn seeding rate recommendations in Indiana, please see: https://www.agry.purdue.edu/ext/corn/news/timeless/PlantPopulations.html

CORN RESPONSE TO PLANTING DATE, HYBRID MATURITY, AND FUNGICIDE (PPAC)

Daniel Quinn: Department of Agronomy, Purdue University
Gary Tragesser: Pinney Purdue Agricultural Center
Alex Leman: Pinney Purdue Agricultural Center
Darcy Telenko: Department of Botany and Plant Pathology, Purdue University

Study Location: Pinney Purdue Agricultural Center, Wanatah, IN
Soil Type: Sebewa loam
Planting Date: 1st Planting Date—May 24, 2022; 2nd Planting Date—June 3, 2022.
Harvest Date: Oct. 27, 2022
Corn Hybrid(s): Pioneer P1099Q and Pioneer P9608Q
Corn Seeding Rate: 30,000 seeds/ac
Corn Nitrogen (N) Fertilizer Rate and Source: 212 lbs N per acre, UAN (28-0-0)
Corn Fungicide Product and Rate Used: Veltyma, 10 oz/ac, applied at the R1 growth stage
Previous Crop: Corn | **Tillage:** Conventional
Study Replications: 4

RESEARCH TRIAL OVERVIEW:

A field research trial was established at the Pinney Purdue Agricultural Center (PPAC) in Porter County, IN. The research trial examined corn yield response to hybrid maturity, planting date, and foliar fungicide application applied at the R1 growth stage (tassel emergence). The trial was designed as a randomized complete block design with four replications. Plots measured 30 feet wide (12, 30-inch corn rows) by 500 feet long, and the center six rows were harvested with a commercial combine containing a calibrated yield monitor and adjusted to 15.5% moisture for yield analysis.

RESULTS:

TABLE 7. *Corn grain moisture and yield in response to hybrid, planting date, and fungicide. Wanatah, IN 2022.*

CORN HYBRID	PLANTING DATE	R1 FUNGICIDE APPLICATION	R1 GROWTH STAGE TIMING	GRAIN MOISTURE %	GRAIN YIELD BU/AC
Pioneer	24-May	Yes	26-July	15.7 e*	209.7 c
'9608Q'	24-May	No	26-July	15.9 e	208.9 c
	3-June	Yes	30-July	17.9 d	217.1 b
	3-June	No	30-July	17.1 d	214.7 bc
Pioneer	24-May	Yes	4-Aug	19.5 c	230.5 a
'1099Q'	24-May	No	4-Aug	19.2 c	227.8 a
	3-June	Yes	12-Aug	24.1 a	225.5 a
	3-June	No	12-Aug	22.6 b	227.3 a
P–value				0.001	0.001

* Mean values that do not contain the same corresponding letter are determined statistically different ($P < 0.1$).

SUMMARY (TAKE-HOME POINTS):

- When comparing yield differences between the two hybrids selected and across both planting dates and fungicide applications, hybrid P1099Q (110-d) outyielded hybrid P9608Q (96-d) by an average of 15 bushels per acre (bu/ac) (Table 7).
- Fungicide application at the R1 growth stage (tassel emergence) did not impact corn grain yield, regardless of hybrid maturity and/or planting date. Lack of fungicide response was likely due to minimal foliar disease observed at this location during the 2022 growing season. Foliar disease was found late in the growing season (R4–R5 growth stage) at low levels and was determined not to be yield limiting.
- The two corn hybrids examined did differ in grain yield response to planting date. The early maturing corn hybrid (P9608Q) had a yield increase with a later planting date. Corn planted on June 3 resulted in an average yield increase of 7 bu/ac for hybrid P9608Q as compared to a May 24 planting date. However, corn hybrid P1099Q did not observe a grain yield difference between the two planting dates.
- Corn grain moisture at harvest was impacted by planting date and not fungicide application. For both corn hybrids examined, a later planting date resulted in a grain moisture increase of 2–3%.
- Overall, the presented results show the ability of certain corn hybrids to not lose yield despite a later planting date; however, grain moisture at harvest was increased. In addition, the earlier maturing corn hybrid actually saw a corn yield increase when planted later. Lastly, an R1 fungicide application failed to increase corn yield at this location in 2022 due to a lack of foliar disease presence.

DAVIS PURDUE AGRICULTURAL CENTER (DPAC)

CORN RESPONSE TO NITROGEN FERTILIZER APPLICATION TIMING FOLLOWING A RYE COVER CROP (DPAC)

Daniel Quinn: Department of Agronomy, Purdue University
Riley Seavers: Department of Agronomy, Purdue University
Shalamar Armstrong: Department of Agronomy, Purdue University
Jeff Boyer: Davis Purdue Agricultural Center, Purdue University

Study Location: Davis Purdue Agricultural Center, Farmland, IN
Soil Type: Pewamo silty clay loam, (0–1% slope), Blount silt loam, ground moraine (0–2% slope)
Planting Date: May 23, 2022 | **Harvest Date:** October 24, 2022
Corn Hybrid: Pioneer P1108Q | **Corn Seeding Rate:** 30,000 seeds per acre
Corn Nitrogen (N) Fertilizer Rate and Source: 200 lbs N per acre, UAN (28-0-0)
Previous Crop: Soybean | **Tillage:** No-Till
Study Replications: 4

RESEARCH TRIAL OVERVIEW:

This research trial was established at the Davis Purdue Agricultural Center in Randolph County, IN. This research trial assessed corn yield differences when utilizing different in-season N fertilizer sidedress application timings following a fall-planted cereal rye cover crop and no rye cover crop. The rye cover crop was fall drill-seeded at a rate of 45 lbs/ac and was chemically terminated with glyphosate two weeks before corn planting. All plots received a 2x2 starter nitrogen application at planting of 40 lbs N/ac. Plots measured 30 feet wide (12, 30-inch corn rows) x 700+ feet long. The center eight rows were harvested with a commercial combine containing a calibrated yield monitor and were adjusted to 15.5% moisture for yield analysis.

RESEARCH TREATMENTS:

1. No cover crop + V5 sidedress N (coulter-inject, UAN 28-0-0)
2. No cover crop + V10 sidedress N (surface-banded, UAN 28-0-0)

3. No cover crop + V5 + V10 sidedress N (60% remaining N at V5 and 40% remaining N at V10)
4. Cereal rye cover crop + V5 sidedress N (coulter-inject, UAN 28-0-0)
5. Cereal rye cover crop + V10 sidedress N (surface-banded, UAN 28-0-0)
6. Cereal rye cover crop + V5 + V10 sidedress N (60% remaining N at V5 and 40% remaining N at V10)

RESULTS:

TABLE 8. *Average cereal rye cover crop aboveground biomass and nutrient uptake values. Rye cover crop biomass was sampled immediately prior to spring termination. Farmland, IN.*

BIOMASS	CARBON	NITROGEN	C:N*
-- lbs/ac --	-- lbs/ac --	-- lbs/ac --	-- lbs/ac --
861.1	370	15.3	24:1

* Carbon (C) to nitrogen (N) ratio of aboveground biomass at termination.

TABLE 9. *Mean plant population, grain moisture, and grain yield differences observed across cereal rye cover crop presence and nitrogen (N) fertilizer application timing. Farmland, IN.*

COVER CROP	N TIMING	V5 PLANT POPULATION	GRAIN MOISTURE	GRAIN YIELD
		-- plants/ac --	-- % --	-- bu/ac --
No Cover	0N	28,500 a	17.7 c*	96.4 c
	V5	29,800 a	21 a	217.16 ab
	V10	29,800 a	20.1 ab	218.2 ab
	V5+V10	28,800 a	20.7 a	223.2 ab
Rye Cover Crop	0N	28,500 a	18.5 bc	90.7 c
	V5	29,500 a	20.2 ab	236.6 a
	V10	29,000 a	21.4 a	209.9 b
	V5+V10	28,000 a	20 ab	209.7 b
P-value		*0.934*	*0.008*	*0.001*

* Mean values that do not contain the same corresponding letter are determined statistically different (P < 0.1).

SUMMARY (TAKE-HOME POINTS):

- Rye cover crop biomass at termination averaged 861 lbs/ac, which contained 370 lbs/ac of carbon, 15 lbs/ac of N, and a C:N ratio of 24:1 (Table 8).
- Across all treatments examined, a rye cover crop did not reduce corn yield.
- A significant interaction ($P < 0.001$) was observed between rye cover crop presence and N timing, which indicates optimum N timing did change with cover crop presence at this location. The V10 and V5+V10 N timings lost yield following a rye cover crop, but did not lose yield following no cover crop.
- The V5 N application timing at DPAC had the highest yield value across both cover crop treatments (226 bu/ac) and outyielded the V10 and V5+V10 timings (Table 9).
- Grain moisture at harvest was decreased in the plots containing 0 lbs N/ac, but no differences were observed between N timings and cover crop presence (Table 9).
- Preliminary results suggest a full rate sidedress N application at the V5 growth stage is required to reach maximum yield potential when following a rye cover crop at this location.

CORN YIELD RESPONSE TO IN-SEASON NITROGEN (N) RATES ESTIMATED FROM SATELLITE IMAGERY (DPAC)

Daniel Quinn: Department of Agronomy, Purdue University
Ana Morales-Ona: Department of Agronomy, Purdue University
Jeff Boyer: Davis Purdue Agricultural Center, Purdue University

Study Location: Davis Purdue Agricultural Center. Farmland, IN
Soil Type: Pewamo silty clay loam (0 to 1% slopes) and Glynwood silt loam (1 to 4% slopes)
Planting Date: May 20, 2022 | **Harvest Date:** October 24, 2022
Corn Hybrid: Pioneer P1108Q | **Corn Seeding Rate:** 31,000 seeds/ac
Farmer normal N rate (FNR): 180 lb N/ac (w/o starter N) | **2x2 Starter:** 40 lb N/ac | **Total N rate:** 220 lb N/ac
Previous Crop: Soybean | **Tillage:** Strip-till
Study Replications: 3

RESEARCH TRIAL OVERVIEW:

This NRCS (Natural Resources Conservation Service) funded study examines the feasibility of using satellite imagery to determine maize N status and mid-season optimum N fertilizer rates. Five preplant N fertilizer treatments were established and applied before planting based on the percentage of the farmer's normal N rate (FNR): (1) 40%, (2) 60%, (3) 80%, (4) 100%, and (5) 120% of the FNR. Plots measured 30 feet wide (12, 30-inch corn rows) by the length of the field. Each treatment was replicated three times in a randomized complete block design. Plots were further delineated into shorter sections, "subblocks" equal to the plot width by 200 ft long. Adjacent subblocks (within the same replication) representing the range of all N rates were considered as a "block" and used to develop variable-rate N prescriptions for the sidedress treatments. At growth stage V13, variable-rate N fertilizer prescriptions were developed through identification of the agronomic optimum N fertilizer rate (AONR) of each block based on NDVI from satellite images (PlanetScope multispectral images, 3-m resolution). Sidedress N was surface-banded between the corn rows with a high-clearance sprayer and applied in the form of UAN (28%) in the areas corresponding to the treatments 40, 60, and 80% FNR. The center eight rows were harvested with a commercial combine using a calibrated yield monitor and adjusted to 15.5% moisture for yield analysis.

RESULTS:

TABLE 10. *Nitrogen application timing rates (lbs/ac) and mean grain yield (bu/ac) across different treatments. Farmland, IN.*

TREATMENT	PREPLANT*	STARTER	SIDEDRESS	TOTAL	GRAIN YIELD
	------------------- lbs N/ac ---------------				---- bu/ac ----
40% FNR preplant + sidedress	66	40	84	190	206.4 c[†]_
60% FNR preplant + sidedress	97	40	55	192	212.4 bc
80% FNR preplant + sidedress	126	40	23	189	215.1 ab
100%FNR preplant	155	40	0	195	221.4 a
120%FNR preplant	194	40	0	234	220.4 a
P–value					*0.030*

* Preplant: May 20 (UAN 28%); Starter: May 20 (Liquid mix); Sidedress: July 13, V13 (UAN 28%).

[†] Mean values that do not contain the same corresponding letter are determined statistically different ($P < 0.1$).

SUMMARY (TAKE-HOME POINTS):

- Total N applied across the treatments ranged from 189 to 234 lbs N/ac, with 195 lbs N/ac being the normal total N rate applied by the farmer (FNR; Table 10).
- For the treatments that received sidedress N application (40, 60, and 80% FNR + sidedress), total N applied ranged from 189 to 192 lbs N/ac, while the treatments with no additional sidedress (100 and 120% FNR) ranged from 195 to 234 lbs N/ac.
- Across all treatments examined, the 100% FNR preplant, 120% FNR preplant, and 80% FNR preplant + sidedress resulted in the highest yields observed.
- The total N rate applied when using a sidedress N prescription derived using satellite imagery did not result in large differences from the 100% FNR and did show a minor ability to reduce the total N rate applied. However, yield was lower with treatments containing 40% FNR preplant + sidedress and 60% FNR preplant + sidedress, which suggests the late-season sidedress N application (V13 growth stage) was likely applied too late, thus causing a reduction in grain yield at this location.

NORTHEAST PURDUE AGRICULTURAL CENTER (NEPAC)

CORN RESPONSE TO INPUT-INTENSIVE MANAGEMENT (NEPAC)

Daniel Quinn: Department of Agronomy, Purdue University
Malena Bartaburu: Department of Agronomy, Purdue University
Darcy Telenko: Department of Botany and Plant Pathology, Purdue University
Stephen Boyer: Northeast Purdue Agricultural Center, Purdue University

Study Location: Northeast Purdue Agricultural Center, Columbia City, IN
Soil Type: Mermill loam, Haskins loam, Morley clay loam (6–12% slope), Coesse silty clay loam, Glynwood loam.
Planting Date: May 23, 2022 | **Harvest Date:** Nov. 3, 2022
Corn Hybrid(s): Pioneer P1108Q | **Corn Seeding Rate:** 30,000 and 36,000 seeds/ac
Corn Nitrogen (N) Fertilizer Rate and Source: 200 lbs N per acre, UAN (28-0-0)
Previous Crop: Soybean | **Tillage:** No-till
Study Replications: 3

RESEARCH TRIAL OVERVIEW:

A field research trial was established at the Northeast Purdue Agricultural Center (NEPAC) near Whitley County, IN. The research trial examined corn yield response to different inputs and was utilized to analyze the impact of each one applied individually and in combination (e.g., intensive treatment). The trial was designed as a randomized complete block design with eight treatments and three replications. Plots measured 30 feet wide (12, 30-inch corn rows) by 700+ feet long, and the center six rows were harvested with a commercial combine and adjusted to 15.5% moisture for yield analysis. All plots received a 2x2 starter N application totaling 40 lbs N/ac.

RESEARCH TREATMENTS:

1. Control treatment (C) based on Purdue University seed rate (30K seeds/ac) and nitrogen (N) fertilizer rate recommendations
2. C + surface banded starter (2x2) fungicide (flutriafol, Xyway LFR, 15 oz/ac)

3. C + 20% increase in corn seeding rate (36K seeds/ac)
4. C + sulfur fertilizer [5.2 gallons/ac as ammonium thiosulfate (ATS) at V5 sidedress]
5. C + foliar micronutrients (zinc, manganese, and boron applied at the V6 growth stage)
6. C + late-season N application [starter N (2x2) + V5 sidedress N (60% remaining N rate) + V10-12 growth stage sidedress N surface-banded with drop tubes on a sprayer (40% remaining N rate), total N rate remained the same as other treatments].
7. C + foliar fungicide applied at the R1 growth stage (prothioconazole, trifloxystrobin, fluopyram, Delaro Complete, 10 oz/ac)
8. Intensive treatment: All additional inputs and management practices applied together

RESULTS:

TABLE 11. *Corn grain yield, grain moisture, treatment cost, and net profit differences observed from applied treatments in 2022. Columbia City, IN.*

TREATMENT	GRAIN MOISTURE	GRAIN YIELD	TREATMENT COST†	NET PROFIT†
	--- % ---	-- bu/ac --	-- $/ac --	-- $/ac --
Control	18.6	222.1 d*	253.4	1226.2 b
C + 2x2 Fungicide	19	228.1 bcd	274.3	1244.9 ab
C + Increased Seed	17.9	230.3 bcd	272.7	1260.8 ab
C + Sulfur	19.1	233.5 b	270.7	1284.5 ab
C + Foliar Micro	18.8	231.3 bcd	274.1	1266.4 ab
C + V10-12 SD N	19	222.9 d	261.1	1223.6 b
C + R1 Fungicide	19.5	231.9 bc	277.9	1266.75 ab
Intensive	18.6	249.2 a	356.1	1303.7 a
P–value	*0.106*	*0.004*	–	*0.031*

* Mean values that do not contain the same corresponding letter are determined statistically different ($P < 0.1$).
† Treatment costs were calculated as the combined cost of corn seed, fertilizer cost, chemical input cost, and application cost. Prices were calculated as an average from various local retailers. Net profit was calculated based on average harvest corn grain cash price ($6.66) + average grain yield – treatment costs.

SUMMARY (TAKE-HOME POINTS):

- The intensive management treatment outyielded the control by 27 bu/ac at this location (Table 11). The individual inputs with yield responses included sulfur (+11 bu/ac) and R1 fungicide (+ 9 bu/ac). Both foliar disease and sulfur deficiency were observed at this location, which likely caused the observed yield responses.
- Despite the high application cost, the intensive treatment resulted in a net profit increase compared to the control at this location. This was largely due to high 2022 corn grain prices (e.g., $6.66) and the large yield response observed (+ 27 bu/ac).
- No yield response was observed from the 2x2 starter fungicide, increased seeding rate, foliar micronutrient, or the late-season sidedress N application.
- Preliminary results suggest both sulfur and R1 fungicide have the potential to increase corn yield at this location when conditions conducive for a response are present (e.g., foliar disease presence).

SOUTHEAST PURDUE AGRICULTURAL CENTER (SEPAC)

CORN RESPONSE TO INPUT-INTENSIVE MANAGEMENT (SEPAC -C5)

Daniel Quinn: Department of Agronomy, Purdue University
Malena Bartaburu: Department of Agronomy, Purdue University
Darcy Telenko: Department of Botany and Plant Pathology, Purdue University
Alex Helms: Southeast Purdue Agricultural Center
Joel Wahlman: Southeast Purdue Agricultural Center

Study Location: Field C5, Southeastern Purdue Agricultural Center, Butlerville, IN
Soil Type: Ryker-Muscatatuck silt loams, karst, undulating, eroded
Planting Date: May 11, 2022 | **Harvest Date:** October 6, 2022
Corn Hybrid(s): Pioneer 1136AM
Corn Nitrogen (N) Fertilizer Rate and Source: 210 lbs N per acre, UAN (28-0-0)
Previous Crop: Soybean | **Tillage:** No-till
Study Replications: 4

RESEARCH TRIAL OVERVIEW:

A field research trial was established at the Southeastern Purdue Agricultural Center near Jennings County, IN. The research trial examined corn yield response to different inputs and was utilized to analyze the impact of each one applied individually and in combination (e.g., intensive treatment). The trial was designed as a randomized complete block design with eight treatments and four replications. Plots measured 30 feet wide (12, 30-inch corn rows) by 500+ feet long, and the center six rows were harvested with a commercial combine and adjusted to 15.5% moisture for yield analysis. All plots received a 2x2 starter N application totaling 40 lbs N/ac.

RESEARCH TREATMENTS:

1. Control treatment (C) based on Purdue University seed rate (30K seeds/ac) and nitrogen (N) fertilizer rate recommendations
2. C + surface banded starter (2x2) fungicide (flutriafol, Xyway LFR, 15 oz/ac)

3. C + 20% increase in corn seeding rate (36K seeds/ac)
4. C + sulfur fertilizer [5.2 gallons/ac as ammonium thiosulfate (ATS) at V5 sidedress]
5. C + foliar micronutrients (zinc, manganese, and boron applied at the V6 growth stage)
6. C + late-season N application [starter N (2x2) + V5 sidedress N (60% remaining N rate) + V10-12 growth stage sidedress N surface-banded with drop tubes on a sprayer (40% remaining N rate), total N rate remained the same as other treatments]
7. C + foliar fungicide applied at the R1 growth stage (prothioconazole, trifloxystrobin, fluopyram, Delaro Complete, 10 oz/ac)
8. Intensive treatment: All additional inputs and management practices applied together

RESULTS:

TABLE 12. *Corn grain yield, grain moisture, treatment cost, and net profit differences observed from applied treatments in 2022. Field C5, Butlerville, IN.*

TREATMENT	GRAIN MOISTURE	GRAIN YIELD	TREATMENT COST†	NET PROFIT†
	--- % ---	-- bu/ac --	-- $/ac --	-- $/ac --
Control	21.0 cd*	261.6 c	253.4	1488.8 bc
C + 2x2 Fungicide	21.4 abc	269.8 abc	274.3	1521.6 ab
C + Increased Seed	21.3 bcd	265.3 bc	272.7	1493.9 abc
C + Sulfur	20.9 d	266.9 abc	270.65	1506.7 abc
C + Foliar Micro	21.1 cd	266.6 abc	274.1	1501.4 abc
C + V10-12 SD N	21.3 bcd	269.2 abc	261.1	1531.5 ab
C + R1 Fungicide	21.6 ab	274.5 a	277.9	1550.2 a
Intensive	21.8 a	271.4 ab	356.1	1451.6 c
P–value	0.004	0.077	–	0.030

* Mean values that do not contain the same corresponding letter are determined statistically different ($P < 0.1$).

† Treatment costs were calculated as the combined cost of corn seed, fertilizer cost, chemical input cost, and application cost. Prices were calculated as an average from various local retailers. Net profit was calculated based on average harvest corn grain cash price ($6.66) + average grain yield – treatment costs.

SUMMARY (TAKE-HOME POINTS):

- The intensive treatment outyielded the control by 10 bu/ac at this location in 2022. Individual input responses included only the R1 fungicide (+13 bu/ac) (Table 12). Foliar fungicide yield response was largely driven by reduced foliar disease presence (e.g., tar spot, gray leaf spot, and northern corn leaf blight).
- Despite observed yield increases, the intensive treatment did not result in a higher net profit over the control at this location in 2022. This result highlights the risk of the large yield increases needed to pay for the high input and application costs required with the intensive treatment.
- The R1 fungicide application did result in an increase in net profit as compared to the control (+$62/ac).
- Preliminary results highlight the importance of selecting individual inputs that can be used to increase corn yield and net profit within the specific year and environment, instead of applying every additional input and management practice.

CORN RESPONSE TO INPUT-INTENSIVE MANAGEMENT (SEPAC—D9)

Daniel Quinn: Department of Agronomy, Purdue University

Malena Bartaburu: Department of Agronomy, Purdue University

Darcy Telenko: Department of Botany and Plant Pathology, Purdue University

Alex Helms: Southeast Purdue Agricultural Center

Joel Wahlman: Southeast Purdue Agricultural Center

Study Location: Field D9, Southeastern Purdue Agricultural Center, Butlerville, IN

Soil Type: Avonburg silt loam (0–2% slope), Cobbsfork silt loam (0–1% slope)

Planting Date: May 10, 2022 | **Harvest Date:** October 14, 2022

Corn Hybrid(s): Dekalb DKC67-44

Corn Nitrogen (N) Fertilizer Rate and Source: 210 lbs N per acre, UAN (28-0-0)

Previous Crop: Soybean | **Tillage:** No-till

Study Replications: 4

RESEARCH TRIAL OVERVIEW:

A field research trial was established at the Southeastern Purdue Agricultural Center, Butlerville, IN. The research trial examined corn yield response to different inputs and was utilized to analyze the impact of each one individually and in combination as an intensive management. The trial was designed as a randomized complete block design with eight treatments and four replications. Plots measured 30 feet wide (12, 30-inch corn rows) by 500+ feet long, and the center six rows were harvested with a commercial combine and adjusted to 15.5% moisture for yield analysis. All plots received a 2x2 starter N application totaling 40 lbs N/ac.

RESEARCH TREATMENTS:

1. Control treatment (C) based on Purdue University seed rate (30K seeds/ac) and nitrogen (N) fertilizer rate recommendations
2. C + surface banded starter (2x2) fungicide (flutriafol, Xyway LFR, 15 oz/ac)
3. C + 20% increase in corn seeding rate (36K seeds/ac)
4. C + sulfur fertilizer [5.2 gallons/ac as ammonium thiosulfate (ATS) at V_5 sidedress]
5. C + foliar micronutrients (zinc, manganese, and boron applied at the V6 growth stage)
6. C + late-season N application [starter N (2x2) + V_5 sidedress N (60% remaining N rate) + V_{10-12} growth stage sidedress N surface-banded with drop tubes on a sprayer (40% remaining N rate), total N rate remained the same as other treatments].
7. C + foliar fungicide applied at the R1 growth stage (prothioconazole, trifloxystrobin, fluopyram, Delaro Complete, 10 oz/ac)
8. Intensive treatment: All additional inputs and management practices applied together

RESULTS:

TABLE 13. *Corn grain yield, grain moisture, treatment cost, and net profit differences observed from applied treatments in 2022. Field D9, Butlerville, IN.*

TREATMENT	GRAIN MOISTURE	GRAIN YIELD	TREATMENT COST[†]	NET PROFIT[†]
	--- % ---	-- bu/ac --	-- $/ac --	-- $/ac --
Control	18.7 c*	261.3 bcd	253.4	1486.9 abc
C + 2x2 Fungicide	18.9 bc	265.9 abc	274.3	1496.9 ab
C + Increased Seed	18.8 bc	258.0 d	272.7	1445.7 c
C + Sulfur	18.9 bc	266.1 abc	270.65	1501.5 a
C + Foliar Micro	18.9 ab	259.6 cd	274.1	1454.7 bc
C + V10-12 SD N	18.8 bc	265.8 abc	261.1	1509.2 a
C + R1 Fungicide	19.2 a	268.0 ab	277.9	1506.9 a
Intensive	19.0 ab	270.4 a	356.1	1444.9 c
P-value	0.025	0.001	–	0.001

* Mean values that do not contain the same corresponding letter are determined statistically different ($P < 0.1$).

† Treatment costs were calculated as the combined cost of corn seed, fertilizer cost, chemical input cost, and application cost. Prices were calculated as an average from various local retailers. Net profit was calculated based on average harvest corn grain cash price ($6.66) + average grain yield – treatment costs.

SUMMARY (TAKE-HOME POINTS):

- The intensive treatment outyielded the control at this location by 9 bu/ac (Table 13). Across individual input comparisons, the R1 fungicide application resulted in the highest yield achieved and likely drove the observed yield increase from the intensive treatment. Fungicide response was likely due to decreased foliar disease presence (e.g., tar spot, gray leaf spot, and northern corn leaf blight).

- Despite observed yield increases, across all treatments examined, no treatment resulted in an increase in overall net profit. This result again highlights the importance of selecting specific inputs based on observed stresses and yield-limiting factors in individual years and environments to maximize both yield and net profit.

- No yield increases were observed from the application of 2x2 starter fungicide, increased seeding rate, sulfur, foliar micronutrient, and late-season sidedress N at this individual location in 2022.

CORN RESPONSE TO NITROGEN FERTILIZER APPLICATION TIMING FOLLOWING A RYE COVER CROP (SEPAC)

Daniel Quinn: Department of Agronomy, Purdue University
Riley Seavers: Department of Agronomy, Purdue University
Shalamar Armstrong: Department of Agronomy, Purdue University
Alex Helms: Southeast Purdue Agricultural Center
Joel Wahlman: Southeast Purdue Agricultural Center

Study Location: Southeast Purdue Agricultural Center, Butlerville, IN
Soil Type: Cobbsfork silt loam, (0–1% slope)
Planting Date: 4/24/2022 | **Harvest Date:** 10/3/2022
Corn Hybrid: Pioneer P0953AM | **Corn Seeding Rate:** 30,000 seeds per acre
Corn Nitrogen (N) Fertilizer Rate and Source: 210 lbs N per acre, UAN (28-0-0)
Previous Crop: Soybean
Tillage: No-till
Study Replications: 3

RESEARCH TRIAL OVERVIEW:

This research trial was established at the Southeast Purdue Agricultural Center in Jennings County, IN. This research trial assessed corn yield differences when utilizing different in-season N fertilizer sidedress application timings following a fall-planted cereal rye cover crop and no rye cover crop. The rye cover crop was fall drill-seeded at a rate of 45 lbs/ac and was chemically terminated with glyphosate two weeks before corn planting. All plots received a 2x2 starter nitrogen application at planting of 40 lbs N/ac. Plots measured 30 feet wide (12, 30-inch corn rows) x 700+ feet long. The center six rows were harvested with a commercial combine containing a calibrated yield monitor and were adjusted to 15.5% moisture for yield analysis.

RESEARCH TREATMENTS:

1. No cover crop + V5 sidedress N (coulter-inject, UAN 28-0-0)
2. No cover crop + V10 sidedress N (surface-banded, UAN 28-0-0)
3. No cover crop + V5 + V10 sidedress N (60% remaining N at V5 and 40% remaining N at V10)
4. Cereal rye cover crop + V5 sidedress N (coulter-inject, UAN 28-0-0)
5. Cereal rye cover crop + V10 sidedress N (surface-banded, UAN 28-0-0)
6. Cereal rye cover crop + V5 + V10 sidedress N (60% remaining N at V5 and 40% remaining N at V10)

TABLE 14. *Average cereal rye cover crop aboveground biomass and nutrient uptake values. Rye cover crop biomass was sampled immediately prior to spring termination. Butlerville, IN.*

BIOMASS	CARBON	NITROGEN	C:N*
-- lbs/ac --	-- lbs/ac --	-- lbs/ac --	-- lbs/ac --
1337	564	19.4	29:1

* Carbon (C) to nitrogen (N) ratio of aboveground biomass at termination.

TABLE 15. *Mean plant population, grain moisture, and grain yield differences observed across cereal rye cover crop presence and nitrogen (N) fertilizer application timing. Butlerville, IN.*

COVER CROP	N TIMING	V5 PLANT POPULATION	GRAIN MOISTURE	GRAIN YIELD
		-- plants/ac --	-- % --	-- bu/ac --
No Cover	V5	29,667	21.9 ab*	270.5 ab
	V10	30,000	22.1 a	251.5 d
	V5+V10	30,333	21.5 bc	275.9 a
Rye Cover Crop	V5	30,000	21.3 c	264.6 bc
	V10	29,333	22.1 a	238.2 e
	V5+V10	30,333	21.3 c	258.2 de
P-value		*0.401*	*0.082*	*0.015*

* Mean values that do not contain the same corresponding letter are determined statistically different ($P < 0.1$).

SUMMARY (TAKE-HOME POINTS):

- Rye cover crop biomass at termination averaged 1,337 lbs/ac, which contained 564 lbs/ac of carbon, 19.4 lbs/ac of N, and a C:N ratio of 29:1 (Table 14).
- Across all treatments examined, a rye cover crop reduced corn yield by 12 bu/ac. Yield reductions were likely due to observed corn N stress and delayed plant growth.
- A significant interaction ($P < 0.001$) was observed between rye cover crop presence and N timing, which indicates optimum N timing differed with rye cover crop presence at this location.
- The V5 N application timing at SEPAC had the highest yield value across both cover crop treatments (270.5 bu/ac in the no rye cover crop treatments and 264.6 bu/ac in the rye cover crop treatments). At this timing, no yield loss was observed with rye cover crop presence. However, with V10 and V5+V10 sidedress, yield was reduced with rye cover crop presence (Table 15).
- Preliminary results suggest a full rate sidedress N application at the V5 growth stage is required when following a rye cover crop, but may not always be required without a rye cover crop.

THROCKMORTON PURDUE AGRICULTURAL CENTER (TPAC)

CORN EMERGENCE AND YIELD RESPONSE TO CLOSING WHEEL TYPE IN A RYE COVER CROP SYSTEM (TPAC)

Daniel Quinn: Department of Agronomy, Purdue University
Riley Seavers: Department of Agronomy, Purdue University
Pete Illingsworth: Throckmorton Purdue Agricultural Center

Study Location: Throckmorton Purdue Agricultural Center, Lafayette, IN
Soil Type: Drummer silt loam, Octagon silt loam, Throckmorton silt loam, and Toronto-Millbrook complex
Planting Date: June 1, 2022 | **Harvest Date:** November 14, 2022
Corn Hybrid(s): Pioneer P1099Q | **Corn Seeding Rate:** 30,000 seeds per acre
Corn Nitrogen (N) Fertilizer Rate and Source: 212 lbs N per acre, UAN (28-0-0)
Rye Cover Crop: VNS Cereal Rye, fall drill-seeded at 45 lbs per acre, chemically terminated 3 weeks prior to corn planting
Previous Crop: Soybean | **Tillage:** No-till
Study Replications: 4

RESEARCH TRIAL OVERVIEW:

A field research trial was established at the Throckmorton Purdue Agricultural Center (TPAC) in Tippecanoe County, IN. The research trial examined corn emergence and yield response to planter closing wheel type in a no-till system with and without a rye cover crop. The trial was designed as a randomized complete block design with four replications. Plots measured 30 feet wide (12, 30-inch corn rows) by 500+ feet long, and the center six rows were harvested with a commercial combine and adjusted to 15.5% moisture for yield analysis. All treatments received a 2x2 starter application of N fertilizer at planting totaling 40 lbs N/ac.

RESULTS:

TABLE 16. *Corn emergence, final stand, grain moisture, and yield in response to cover crop presence and closing wheel type. Lafayette, IN 2022.*

COVER CROP PRESENCE	CLOSING WHEEL TYPE[*]	%EMERGENCE (6 DAP[*])	FINAL PLANT STAND	GRAIN MOISTURE	GRAIN YIELD
		%	plants/ac	%	bu/ac
No Cover Crop	JD Standard Rubber	33.5 a[†]	26,750 a	18.1 a	206 a
	CA Cruiser Extreme	38.4 a	27,143 a	18.1 a	211 a
	MT Cupped Razor	34.0 a	26,750 a	18.2 a	208 a
Cereal Rye Cover	JD Standard Rubber	35.9 a	25,750 a	18.3 a	209 a
	CA Cruiser Extreme	23.9 a	27,000 a	18.2 a	203 a
	MT Cupped Razor	22.7 a	25,500 a	18.2 a	205 a
P-value		0.215	0.356	0.453	0.534

[*] JD: John Deere, CA: Copperhead Ag, MT: Martin-Till, DAP: days after corn planting

[†] Mean values that do not contain the same corresponding letter are determined statistically different ($P < 0.1$).

FIGURES 5A AND 5B. Photos of the two aftermarket closing wheels, which were tested in comparison to the standard rubber closing wheel: Cruiser Extreme from Copperhead Ag (*left*) and the Cupped Razor from Martin-Till (*right*). Both closing wheels are marketed for high-residue systems and were set to manufacturer specifications.

FIGURES 6A, 6B, AND 6C. Observed differences in furrow closure and seed-to-soil contact between the standard rubber closing wheel (*left*, 6 DAP), the Cruiser Extreme closing wheel (*middle*, 1 DAP), and the Cupped Razor closing wheel (*right*, 1 DAP).

SUMMARY (TAKE-HOME POINTS):

- Across the treatments examined, minimal emergence, plant stand, and yield differences were observed between cover crop presence and closing wheel type (Table 16).
- Corn was planted later than expected (June 1, 2022) due to spring rainfall and was planted into warm soil temperatures and dry soil conditions, which likely caused strong and rapid emergence and stand establishment conditions, and negated any potential differences observed between the examined treatments. In addition, the rye cover crop was successfully terminated and minimal residue issues/interference were observed.
- Visually, both the Cruiser Extreme closing wheel and Cupped Razor closing wheel resulted in better seed-to-soil contact and residue management at planting as compared to the standard rubber closing wheel (Figure 6).
- Across all treatments examined, the Cruiser Extreme closing wheel trended with the highest final plant stand (Table 16). In addition, in a rye cover crop system, the standard rubber closing wheel trended with the fastest plant emergence (Table 16), which was likely due to shallower seed depth and poorer furrow closure, thus driving faster emergence (Table 16). However, no statistical differences were observed. This study will be repeated at more locations in 2023.

EVALUATION OF CORN EMERGENCE AND YIELD RESPONSE TO SEED DEPTH AND HYBRID TYPE (TPAC)

Daniel Quinn: Department of Agronomy, Purdue University

Pete Illingsworth: Throckmorton Purdue Agricultural Center

Tom Bechman: Indiana Prairie Farmer, Franklin, IN 46131

Study Location: Throckmorton Purdue Agricultural Center, Lafayette, IN

Soil Type: Throckmorton silt loam (1–3% slope), Toronto-Millbrook complex (0–2% slope)

Planting Date: May 13, 2022 | **Harvest Date:** November 7, 2022

Corn Hybrid: Becks 6241Q and Becks 5909AM

Corn Seeding Rate: 32,000 seeds per acre

Corn Nitrogen (N) Fertilizer Rate and Source: 210 lbs N per acre, UAN (28-0-0)

Previous Crop: Soybean | **Tillage:** Conventional

Study Replications: 3

RESEARCH TRIAL OVERVIEW:

A field research trial was established at the Throckmorton Purdue Agricultural Center (TPAC) in Tippecanoe County, IN. The research trial examined corn seedling emergence timing and yield differences between four different seeding depths and two different hybrids. The trial was designed as a randomized complete block design with three replications. Plots measured 30 feet wide (12, 30-inch corn rows) by 400 feet long, and the center six rows were harvested with a commercial combine and adjusted to 15.5% moisture for yield analysis.

RESULTS:

TABLE 17. *Mean emergence, population, grain yield, and grain moisture differences observed across treatments. Lafayette, IN.*

CORN HYBRID	SEED PLANTING DEPTH	EMERGENCE 7 DAP*	EMERGENCE 9 DAP	EMERGENCE 12 DAP	FINAL PLANT POPULATION	GRAIN YIELD	GRAIN MOISTURE
		%	%	%	Plants/ac	Bu/ac	%
Becks 6241Q	1 inch	43.7 a[†]	82.3 ab	89.5	28.8K c	218.9 a	16.7 a
	2 inch	5.2 bc	70.8 dc	90.6	29.3K abc	220.2 a	16.7 a
	2.5 inch	1.1 c	66.7 de	89.6	29.2K bc	211.3 b	16.9 a
	3 inch	0 c	46.9 f	88.5	28.7K c	204.5 cd	16.7 a
Becks 5909AM	1 inch	41.7 a	87.5 a	93.8	30.2K ab	199.8 d	15.4 b
	2 inch	15.6 b	84.4 ab	92.7	30.5K a	204.5 cd	15.2 bc
	2.5 inch	2.1 bc	71.9 bcd	92.7	30.2K ab	205.6 cd	15.1 c
	3 inch	2.1 bc	55.2 ef	93.8	30.7K a	205.2 cd	15.4 b
P-value		*0.001*	*0.001*	*0.155*	*0.087*	*0.002*	*0.001*

* Percentage of corn plants emerged; DAP: days after planting

[†] Mean values that do not contain the same corresponding letter are determined statistically different ($P < 0.1$).

SUMMARY (TAKE-HOME POINTS):

- Across all seeding rates examined, the corn hybrid Becks 6241Q (112-d) outyielded the other hybrid in this study, Becks 5909AM (109-d), by an average of 10 bu/ac (214 bu/ac vs. 204 bu/ac) (Table 17).

- Across both hybrids examined, the seeding depth of 1 inch resulted in the fastest emergence (average 42% emerged 7 days after planting), whereas the 3-inch depth resulted in the slowest emergence (average 1% emerged 7 days after planting) (Table 17 and Figure 7). Despite the difference in emergence timing, final emergence percentage and final plant stand was not different between the two hybrids or the four seeding depths examined (Table 17 and Figure 7).

- The two different corn hybrids varied in their responses to planting depth. The hybrid Becks 6241Q had the highest yield at the 1- and 2-inch planting depths, whereas the hybrid Becks 5909AM did not show yield differences across planting depths (Table 17). In addition, corn hybrid Becks 6241Q exhibited a higher tolerance to shallow planting depths in comparison the hybrid Becks 5909AM (Table 17).

- Preliminary results suggest different hybrids may respond differently to various planting depths. However, this study was only performed in one location and in one year. Overall, this research highlights the importance of choosing the correct planting depth, and across both hybrids, a 2-inch planting depth resulted in the highest yield.

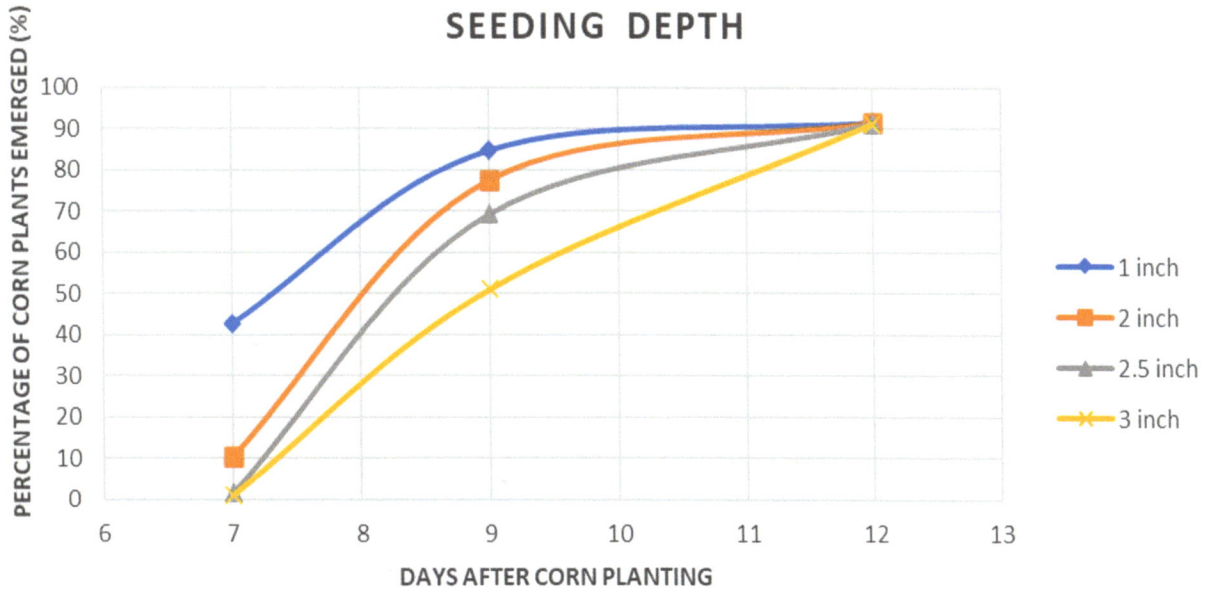

FIGURE 7. Corn seedling emergence (%) after corn planting in response to seeding depth. Data includes both hybrids. Lafayette, IN.

ON-FARM RESEARCH TRIAL RESULTS

CORN YIELD RESPONSE TO IN-SEASON NITROGEN (N) RATES ESTIMATED FROM SATELLITE IMAGERY (WHITE CO.)

Daniel Quinn: Department of Agronomy, Purdue University
Ana Morales-Ona: Department of Agronomy, Purdue University

Study Location: White County, Indiana | **Field ID:** Carter Home
Soil Type: Rensselaer clay loam, Alvin fine sandy loam (0 to 2% and 2 to 6% slope), Pella silty clay loam, and Darroch silt loam
Planting Date: May 12, 2022 | **Harvest Date:** October 8, 2022
Corn Hybrid: Dekalb DKC62-51 | **Corn Seeding Rate:** 32,500 seeds per acre
Farmer's Normal N Rate (FNR): 150 (w/o starter N) | **Starter:** 0 | **Total N rate:** 150 lb/acre
Previous Crop: Soybean | **Tillage:** Conventional
Study Replications: 3

RESEARCH TRIAL OVERVIEW:

This NRCS (Natural Resources Conservation Service) funded study examines the feasibility of using satellite imagery to determine maize N status and mid-season optimum N fertilizer rates. Five early-season N fertilizer treatments were established and applied after planting (V2 growth stage) based on the percentage of the farmer's N rate (FNR): (1) 40%, (2) 60%, (3) 80%, (4) 100%, and (5) 120% of the FNR. Plots were 60 feet wide (24, 30-inch corn rows) by the length of the field. Each treatment was replicated three times in a randomized complete block design. Plots were further delineated into shorter sections, "subblocks," equal to the plot width by 300 ft long. Subblocks (within the same replication) representing the range of all N rates were considered as a "block" and used to determine variable-rate N prescriptions. At growth stage V10, variable-rate N fertilizer prescriptions were developed through identification of the agronomic optimum N fertilizer rate (AONR) of each block based on NDVI from satellite imagery (PlanetScope multispectral images, 3-m resolution). A second sidedress N application was applied in the form of UAN (28%) in the areas corresponding to the treatments of 40, 60, and 80% FNR. Corn yield was harvested with a commercial combine containing a calibrated yield monitor and adjusted to 15.5% moisture for yield analysis.

RESULTS:

TABLE 18. *Mean nitrogen fertilizer application rates (lbs/ac) and mean grain yield (bu/ac) differences observed per treatment. White County, IN.*

TREATMENT	STARTER*	SIDEDRESS 1	SIDEDRESS 2	TOTAL	GRAIN YIELD
		------------ lbs N/ac -----------			----bu/ac----
40% FNR sidedress 1 + sidedress 2	0	60	89	149	214.3 b[†]
60% FNR sidedress 1 + sidedress 2	0	90	58	148	223.2 ab
80% FNR sidedress 1 + sidedress 2	0	121	28	149	224.4 a
100% FNR sidedress	0	150	0	150	218.6 ab
120% FNR sidedress	0	180	0	180	224.1 a
P–value					*0.010*

* Starter: NA; 1st sidedress: June 1, ~V2 (anhydrous NH3); 2nd sidedress: June 28, V10 (UAN 28%)

[†] Mean values that do not contain the same corresponding letter are determined statistically different ($P < 0.1$).

SUMMARY (TAKE-HOME POINTS):

- Total N applied across the treatments ranged from 149 to 180 lbs N/ac, with 150 lbs N/ac being the normal total N rate applied by the farmer (FNR; Table 18).
- For the treatments that received sidedress N application (40, 60, and 80% FNR + sidedress), total N applied ranged within 148 to 149 lbs N/ac. However, yield response to the 40% FNR + sidedress treatment was lower (214.3 bu/ac) compared to the other two sidedress treatments (223.2 and 224.4 bu/ac).
- Across all treatments examined, the 100% FNR, 120% FNR, 60% FNR + sidedress, and 80% FNR + sidedress resulted in the highest yields observed.
- The total N rate applied when using a variable-rate sidedress N prescription derived using satellite imagery did not result in large differences from the 100% FNR. However, yield was lower with treatments containing 40% FNR preplant + sidedress, which suggests the higher late-season sidedress N application (89 lbs N/ac, V10 growth stage) for this specific treatment may have been too late, thus causing a reduction in grain yield at this location.

CORN YIELD RESPONSE TO IN-SEASON NITROGEN (N) RATES ESTIMATED FROM SATELLITE IMAGERY (WHITE CO.)

Daniel Quinn: Department of Agronomy, Purdue University

Ana Morales-Ona: Department of Agronomy, Purdue University

Study Location: White County, Indiana | **Field ID:** Don209 (Hybrid A)

Soil Type: Mundelein silt loam (0 to 2% slopes) and Pella silty clay loam

Planting Date: May 16, 2022 | **Harvest Date:** October 15, 2022

Corn Hybrid: Dekalb DKC 62-04 | **Corn Seeding Rate:** 32,500 seeds per acre

Farmer's Normal N Rate (FNR): 150 (w/o starter N) | **Starter:** 0 | **Total N rate:** 150 lbs N/ac

Previous Crop: Soybean | **Tillage:** Conventional

Study Replications: 3

RESEARCH TRIAL OVERVIEW:

This NRCS (Natural Resources Conservation Service) funded study examines the feasibility of using satellite imagery to determine maize N status and mid-season optimum N fertilizer rates. Five preplant N fertilizer treatments were established and applied after planting (V2 growth stage) based on the percentage of the farmer's N rate (FNR): (1) 40%, (2) 60%, (3) 80%, (4) 100%, and (5) 120% of the FNR. Plots were 60 feet wide (24, 30-inch corn rows) by the length of the field. Each treatment was replicated three times in a randomized complete block design. Plots were further delineated into shorter sections, "subblocks," equal to the plot width by 300 ft long. Subblocks (within the same replication) representing the range of all N rates were considered as a "block" and used to determine variable-rate N prescriptions. At growth stage V9, variable-rate N fertilizer prescriptions were developed through identification of the agronomic optimum N fertilizer rate (AONR) of each block based on NDVI from satellite (PlanetScope multispectral images, 3-m resolution). A second sidedress N was applied in the form of UAN (28%) in the areas corresponding to the treatments of 40, 60, and 80% FNR. The field was harvested with a commercial combine containing a calibrated yield monitor and adjusted to 15.5% moisture for yield analysis.

RESULTS:

TABLE 19. *Mean nitrogen fertilizer application rates (lbs/ac) and mean grain yield (bu/ac) differences observed per treatment. White County, IN.*

TREATMENT	STARTER*	SIDEDRESS 1	SIDEDRESS 2	TOTAL	GRAIN YIELD
		------------ lbs N/ac -----------			----bu/ac----
40% FNR sidedress 1 + sidedress 2	0	60	89	149	222.6 a[†]_
60% FNR sidedress 1 + sidedress 2	0	90	62	152	212.4 b
80% FNR sidedress 1 + sidedress 2	0	119	32	151	224.3 a
100% FNR sidedress	0	148	0	148	215.9 ab
120% FNR sidedress	0	179	0	179	224.3 a
P-value					*0.001*

* Starter: NA; 1st sidedress: June 2, ~V2 (NH3); 2nd sidedress: June 28, V9 (UAN 28%)

[†] Mean values that do not contain the same corresponding letter are determined statistically different ($P < 0.1$).

SUMMARY (TAKE-HOME POINTS):

- Total N applied across the treatments ranged from 148 to 179 lbs N/ac, with 148 lbs N/ac being the normal total N rate applied by the farmer (FNR; Table 19).
- For the treatments that received sidedress N application (40, 60, and 80% FNR + sidedress), total N applied ranged within 148 to 152 lbs N/ac.
- Across all treatments examined, the 100% FNR, 120% FNR, 40% FNR + sidedress, and 80% FNR + sidedress resulted in the highest yields observed.
- The total N rate applied when using a sidedress N prescription derived using satellite imagery did not result in large differences from the 100% FNR. However, yield was lower with treatments containing 60% FNR preplant + sidedress. It is unclear as to why the 40% FNR + sidedress and the 60% FNR sidedress differed in observed grain yields.

CORN YIELD RESPONSE TO IN-SEASON NITROGEN (N) RATES ESTIMATED FROM SATELLITE IMAGERY (DUBOIS CO.)

Daniel Quinn: Department of Agronomy, Purdue University
Ana Morales-Ona: Department of Agronomy, Purdue University

Study Location: Dubois County, Indiana
Soil Type: Peoga silt loam and Dubois silt loam (0 to 2% slopes)
Planting Date: May 1, 2022 | **Harvest Date:** Sept. 17, 2022
Corn Hybrid: Pioneer P1718AML | **Corn Seeding Rate:** 32,000 seeds per acre
Farmer's Normal N Rate (FNR): 150 (w/o starter N) | **Starter:** 52 | **Total N rate:** 202 lbs N/ac
Previous Crop: Wheat (soybean-wheat-corn rotation) | **Tillage:** No-till
Study Replications: 4

RESEARCH TRIAL OVERVIEW:

This NRCS (Natural Resources Conservation Service) funded study examines the feasibility of using satellite imagery to determine maize N status and mid-season optimum N fertilizer rates. Five preplant N fertilizer treatments were established and applied before planting based on the percentage of the farmer's N rate (FNR): (1) 40%, (2) 60%, (3) 80%, (4) 100%, and (5) 120% of the FNR. Plots were 80 feet wide (36, 30-inch corn rows) by the length of the field. Each treatment was replicated four times in a randomized complete block design. Plots were further delineated into shorter sections, "subblocks," equal to the plot width by 200 ft long. Subblocks (within the same replication) representing the range of all N rates were considered as a "block" and used to determine variable-rate N prescriptions. At growth stage V7, variable-rate N fertilizer prescriptions were developed through identification of the agronomic optimum N fertilizer rate (AONR) of each block based on NDVI from satellite (PlanetScope multispectral images, 3-m resolution). Sidedress N was applied in the form of urea in the areas corresponding to the treatments of 40, 60, and 80% FNR. The field was harvested with a commercial combine containing a calibrated yield monitor and adjusted to 15.5% moisture for yield analysis.

RESULTS:

TABLE 20. *Mean nitrogen fertilizer application rates (lbs/ac) and mean grain yield (bu/ac) differences observed per treatment. Dubois County, IN.*

TREATMENT	PREPLANT*	STARTER	SIDEDRESS	TOTAL	GRAIN YIELD
	----------- lbs N/ac -----------				---- bu/ac ----
40% FNR preplant + sidedress	63	52	52	167	225.2 a[†]
60% FNR preplant + sidedress	91	52	23	166	220.2 a
80% FNR preplant + sidedress	126	52	7	185	230.6 a
100% FNR preplant	154	52	0	206	223.6 a
120% FNR preplant	180	52	0	232	222.6 a
P-value					0.220

* Preplant: March 17 (anhydrous NH3); Starter: May 1 (Liquid mix); Sidedress: June 11, V7–V8 (Urea + stabilizer)

[†] Mean values that do not contain the same corresponding letter are determined statistically different ($P < 0.1$).

SUMMARY (TAKE-HOME POINTS):

- Total N applied across the treatments ranged from 166 to 232 lbs N/ac with 206 lbs N/ac being the normal total N rate applied by the farmer (FNR; Table 20).

- For the treatments that received sidedress N application (40, 60, and 80%FNR + sidedress), total N applied ranged from 166 to 185 lbs N/ac, while for the treatments with no additional sidedress (100 and 120% FNR), total N applied ranged from 206 to 232 lbs N/ac.

- Across all treatments examined, no yield differences were observed. However, preliminary results at this location shows the potential of a variable-rate sidedress N application prescription developed from satellite imagery to reduce overall total N rate applied while also maintaining yield potential in comparison to the 100% FNR treatment.

CORN YIELD RESPONSE TO IN-SEASON NITROGEN (N) RATES ESTIMATED FROM SATELLITE IMAGERY (MARSHALL CO.)

Daniel Quinn: Department of Agronomy, Purdue University

Ana Morales-Ona: Department of Agronomy, Purdue University

Study Location: Marshall County, Indiana

Soil Type: Brookston loam (0–1% slope), Crosier loam (0–1% slope), and Selfridge-Crosier complex (0–1% slope)

Planting Date: May 19, 2022 | **Harvest Date:** October 16, 2022

Corn Hybrid: Dekalb DKC61-41RIB | **Corn Seeding Rate:** 32,000 seeds per acre

Farmer's Normal N Rate (FNR): 150 (w/o starter N) | **Starter:** 30 | **Total N rate:** 180 lbs N/ac

Previous Crop: Soybean | **Tillage:** Conventional

Study Replications: 4

RESEARCH TRIAL OVERVIEW:

This NRCS (Natural Resources Conservation Service) funded study examines the feasibility of using satellite imagery to determine maize N status and mid-season optimum N fertilizer rates. Five preplant N fertilizer treatments were established and applied before planting based on the percentage of the farmer's N rate (FNR): (1) 40%, (2) 60%, (3) 80%, (4) 100%, and (5) 120% of the FNR. Plots measured 40 feet wide (16, 30-inch corn rows) by the length of the field. Each treatment was replicated four times in a randomized complete block design. Plots were further delineated into shorter sections, "subblocks," equal to the plot width by 200 ft long. Adjacent subblocks (within the same replication) representing the range of all N rates were considered as a "block" and used to determine variable-rate N prescriptions for sidedress treatments. At growth stage V7, variable-rate N fertilizer prescriptions were developed through identification of the agronomic optimum N fertilizer rate (AONR) of each block based on NDVI from satellite (PlanetScope multispectral images, 3-m resolution). Sidedress N was applied in the form of UAN (28%) in the areas corresponding to the treatments of 40, 60, and 80% FNR. The center 8 rows were harvested with a commercial combine and adjusted to 15.5% moisture for yield analysis.

RESULTS:

TABLE 21. *Mean nitrogen fertilizer application rates (lbs/ac) and mean grain yield (bu/ac) differences observed per treatment. Marshall County, IN.*

TREATMENT	PREPLANT*	STARTER	SIDEDRESS	TOTAL	GRAIN YIELD
	------------ lbs N/ac -----------				---- bu/ac ----
40% FNR preplant + sidedress	60	30	68	158	232.7 b[†]
60% FNR preplant + sidedress	90	30	37	157	234.4 b
80% FNR preplant + sidedress	120	30	11	161	233.3 b
100% FNR preplant	150	30	0	180	235.1 ab
120% FNR preplant	180	30	0	210	240.9 ab
P-value					0.020

* Preplant: March 17 (UAN 28%); Starter: May 19 (Liquid mix); Sidedress: June 21, V7–V8 (UAN 28%)

[†] Mean values that do not contain the same corresponding letter are determined statistically different ($P < 0.1$).

SUMMARY (TAKE-HOME POINTS):

- Total N applied across the treatments ranged from 158 to 210 lbs N/ac, with 180 lbs N/ac being the normal total N rate applied by the farmer (FNR; Table 21).
- For the treatments that received sidedress N application (40, 60, and 80% FNR + sidedress), total N applied ranged from 157 to 161 lbs N/ac, while for the treatments with no additional sidedress (100 and 120% FNR), total N applied ranged from 180 to 210 lbs N/ac.
- Despite the difference (~20 lbs./acre) in the total N applied between the normal total N rate (180 lbs./acre) and the treatments with sidedress N (157 to 161 lbs./acre), grain yield was not statistically different.
- Across all treatments examined, no yield differences were observed. However, preliminary results at this location show the ability of a sidedress N application prescription developed from satellite imagery can reduce overall total N rate applied while also maintaining yield potential in comparison to the 100% FNR treatment.

CORN YIELD RESPONSE TO IN-SEASON NITROGEN (N) RATES ESTIMATED FROM SATELLITE IMAGERY (WHITE CO.)

Daniel Quinn: Department of Agronomy, Purdue University
Ana Morales-Ona: Department of Agronomy, Purdue University

Study Location: White County, Indiana | **Field ID:** Griner
Soil Type: Pella silty clay loam and Mundelein silt loam (0 to 2% slopes)
Planting Date: May 13, 2022 | **Harvest Date:** October 17, 2022
Corn Hybrid: DKC64-30W | **Corn Seeding Rate:** 34,000 seeds per acre
Farmer's Normal N Rate (FNR): 150 (w/o starter N) | **Starter:** 0 | **Total N rate:** 150 lbs./acre
Previous Crop: Corn | **Tillage:** Conventional
Study Replications: 3

RESEARCH TRIAL OVERVIEW:

This NRCS (Natural Resources Conservation Service) funded study examines the feasibility of using satellite imagery to determine maize N status and mid-season optimum N fertilizer rates. Five postplant N fertilizer treatments were established and applied after planting (V2 growth stage) based on the percentage of the farmer's N rate (FNR): (1) 40%, (2) 60%, (3) 80%, (4) 100%, and (5) 120% of the FNR. Plots were 60 feet wide (24, 30-inch corn rows) by the length of the field. Each treatment was replicated three times in a randomized complete block design. Plots were further delineated into shorter sections, "subblocks," equal to the plot width by 300 ft long. Subblocks (within the same replication) representing the range of all N rates were considered as a "block" and used to develop variable-rate N prescriptions for the sidedress treatments. At growth stage V9, variable-rate N fertilizer prescriptions were developed through identification of the agronomic optimum N fertilizer rate (AONR) of each block based on NDVI from satellite (PlanetScope multispectral images, 3-m resolution). A second sidedress N was applied in the form of UAN (28%) in the areas corresponding to the treatments of 40, 60, and 80% FNR. Grain yield was determined with a commercial combine containing a calibrated yield monitor and adjusted to 15.5% moisture for yield analysis.

RESULTS:

TABLE 22. *Mean nitrogen fertilizer application rates (lbs/ac) and mean grain yield (bu/ac) differences observed per treatment. White County, IN.*

TREATMENT	STARTER*	SIDEDRESS 1	SIDEDRESS 2	TOTAL	GRAIN YIELD
		------------ lbs N/ac -----------			----bu/ac----
40% FNR sidedress 1 + sidedress 2	0	72	101	173	231.3 bc[†]
60% FNR sidedress 1 + sidedress 2	0	108	65	173	230.3 c
80% FNR sidedress 1 + sidedress 2	0	143	29	172	230.2 c
100% FNR sidedress	0	179	0	179	235.7 ab
120% FNR sidedress	0	215	0	215	235.5 a
P-value					*0.010*

* Starter: NA; 1st sidedress: June 1, ~V2 (NH3); 2nd sidedress: June 28, V9 (UAN 28%)

[†] Mean values that do not contain the same corresponding letter are determined statistically different ($P < 0.1$).

SUMMARY (TAKE-HOME POINTS):

- Total N applied across the treatments ranged from 172 to 215 lbs N/ac, with 179 lbs N/ac being the normal total N rate applied by the farmer (FNR; Table 22).
- For the treatments that received sidedress N application (40, 60, and 80% FNR + sidedress), total N applied ranged within 172 and 173 lbs N/ac, while for the treatments with no additional sidedress (100 and 120% FNR), total N applied ranged from 179 to 215 lbs N/ac.
- Across all treatments examined, the 100% FNR and 120% FNR treatments resulted in the highest yields observed.
- The total N rate applied when using a sidedress N prescription derived using satellite imagery was slightly lower when compared to the 100% FNR. However, yield was lower with treatments containing 60% FNR + sidedress and 80% FNR + sidedress, which suggests the lower N rate applied combined with the later in-season sidedress N application for these specific treatments may have been too late, thus causing a reduction in grain yield at this location.

CORN YIELD RESPONSE TO IN-SEASON NITROGEN (N) RATES ESTIMATED FROM SATELLITE IMAGERY (CLAY CO.)

Daniel Quinn: Department of Agronomy, Purdue University

Ana Morales-Ona: Department of Agronomy, Purdue University

Study Location: Clay County, Indiana

Soil Type: Evansville silt loam (occasionally flooded), Bonnie silt loam (frequently flooded)

Planting Date: May 10, 2022 | **Harvest Date:** Sept. 22, 2022

Corn Hybrid: DKC-6744 | **Corn Seeding Rate:** 32,500 seeds per acre

Farmer's Normal N Rate (FNR): 160 (w/o starter N) | **Starter:** 35 | **Total N Rate:** 195 lb/acre

Previous Crop: Soybean | **Tillage:** Conventional

Study Replications: 3

RESEARCH TRIAL OVERVIEW:

This NRCS (Natural Resources Conservation Service) funded study examines the feasibility of using satellite imagery to determine maize N status and mid-season optimum N fertilizer rates. Five preplant N fertilizer treatments were established and applied before planting based on the percentage of the farmer's N rate (FNR): (1) 40%, (2) 60%, (3) 80%, (4) 100%, and (5) 120% of the FNR. Plots were 60 feet wide (24, 30-inch corn rows) by the length of the field. Each treatment was replicated three times in a randomized complete block design. Plots were further delineated into shorter sections, "subblocks," equal to the plot width by 200 ft long. Subblocks (within the same replication) representing the range of all N rates were considered as a "block." At growth stage V8, variable-rate N fertilizer prescriptions were developed through identification of the agronomic optimum N fertilizer rate (AONR) of each block based on NDVI from satellite (PlanetScope multispectral images, 3-m resolution). Sidedress N was applied in the form of UAN (28%) in the areas corresponding to the treatments of 40, 60, and 80% FNR. The field was harvested with a commercial combine and adjusted to 15.5% moisture for yield analysis.

RESULTS:

TABLE 23. *Mean nitrogen fertilizer application rates (lbs/ac) and mean grain yield (bu/ac) differences observed per treatment. Clay County, IN.*

TREATMENT	PREPLANT*	STARTER	SIDEDRESS	TOTAL	GRAIN YIELD
	------------ lbs N/ac -----------				---- bu/ac ----
40% FNR preplant + sidedress	71	35	83	189	291.7 a[†]
60% FNR preplant + sidedress	95	35	40	170	292.5 a
80% FNR preplant + sidedress	127	35	8	170	291.9 a
100% FNR preplant	161	35	0	196	299.2 a
120% FNR preplant	187	35	0	222	294.9 a
P-value					0.260

* Preplant: April 28 (NH_3); Starter: May 10 (Liquid mix); Sidedress: June 17 (~V8–V9) (UAN 28%).

[†] Mean values that do not contain the same corresponding letter are determined statistically different ($P < 0.1$).

SUMMARY (TAKE-HOME POINTS):

- Total N applied across the treatments ranged from 189 to 222 lbs N/ac, with 196 lbs N/ac being the normal total N rate applied by the farmer (FNR; Table 23)
- For the treatments that received sidedress N application (40, 60, and 80% FNR + sidedress), total N applied ranged from 170 to 189 lbs N/ac, while for the treatments with no additional sidedress (100 and 120% FNR), total N applied ranged from 196 to 222 lbs N/ac.
- Across all treatments examined, no statistical differences were observed. The results from this location showcase the ability of total N rate predictions produced from satellite imagery have the potential to be lower than what is currently being applied by the farmer, while also maintaining optimum grain yield.

APPENDIX—WEATHER DATA

TABLE 24. *Mean monthly environmental conditions for the Purdue Agronomy Center for Research and Education (ACRE), Pinney Purdue Agricultural Center (PPAC), Davis Purdue Agricultural Center (DPAC), Northeast Purdue Agricultural Center (NEPAC), Southeast Purdue Agricultural Center (SEPAC), and Throckmorton Purdue Agricultural Center (TPAC). Indiana, 2022.*

MONTH	ACRE			PPAC			DPAC		
	TEMP. MIN.† °F	TEMP. MAX.† °F	TOTAL PRECIP. (IN)	TEMP. MIN. °F	TEMP. MAX. °F	TOTAL PRECIP. (IN)	TEMP. MIN. °F	TEMP. MAX. °F	TOTAL PRECIP. (IN)
January	12.5	32.4	0.47	9.8	27.5	0.25	12.0	32.1	1.66
February	18.3	37.5	2.28	16.9	33.8	1.72	19.6	38.6	3.13
March	32.8	52.8	3.40	29.2	48.3	2.86	31.8	53.2	4.05
April	39.6	59.8	2.74	35.1	54.6	3.09	38.4	58.2	2.81
May	55.5	75.8	5.77	51.5	71.9	2.72	53.8	74.8	3.63
June	60.0	84.4	1.20	57.2	80.9	2.11	58.8	84.7	1.33
July	64.0	84.5	1.74	61.2	81.4	3.58	64.4	84.4	5.61
August	61.0	83.3	4.47	58.7	80.4	3.55	60.1	82.7	2.89
September	53.1	77.5	1.80	51.3	74.2	1.34	52.8	76.4	1.85
October	40.4	65.7	2.73	38.2	62.7	4.09	39.2	65.7	0.87
November	33.4	52.8	1.97	30.8	49.7	1.18	31.7	53.9	0.86
December	24.5	38.4	1.25	20.9	34.2	1.06	22.9	38.5	1.82
Annual	*41.4*	*62.2*	*29.82*	*38.5*	*58.4*	*27.55*	*40.6*	*62.0*	*30.51*

MONTH	NEPAC			SEPAC			TPAC		
	TEMP. MIN. °F	TEMP. MAX. °F	TOTAL PRECIP. (IN)	TEMP. MIN. °F	TEMP. MAX. °F	TOTAL PRECIP. (IN)	TEMP. MIN. °F	TEMP. MAX. °F	TOTAL PRECIP. (IN)
January	11.5	29.8	0.51	18.2	37.1	2.61	14.3	33.2	0.55
February	17.9	35.2	2.61	23.8	44.8	6.43	18.5	37.8	2.48
March	31.7	51.6	3.93	35.2	58.2	3.60	33.7	53.3	3.56
April	37.8	56.2	3.49	42.0	62.9	3.57	40.5	60.4	2.05
May	53.6	73.2	4.40	55.5	77.6	5.04	56.5	76.3	5.17
June	59.9	83.1	1.65	60.4	85.9	3.60	61.8	85.8	0.61
July	63.6	82.8	8.06	67.1	86.6	7.04	65.8	85.6	1.84
August	60.9	82.0	2.31	62.9	85.1	4.38	62.9	84.1	5.58
September	53.7	75.5	1.39	56.5	80.4	3.54	55.0	78.7	0.72
October	40.1	64.3	2.86	40.7	68.4	1.50	41.7	66.6	2.26
November	32.8	51.6	2.30	35.0	57.9	1.16	34.1	53.4	1.39
December	23.5	35.7	2.04	26.7	43.3	2.65	26.7	39.1	1.66
Annual	*40.8*	*60.2*	*35.55*	*43.8*	*65.8*	*45.12*	*42.6*	*62.9*	*27.88*

*Temperature and precipitation data acquired from the Indiana State Climate Office and the Purdue mesonet stations. https://ag.purdue.edu/indiana-state-climate/purdue-mesonet/purdue-mesonet-data-hub/

† Average minimum and maximum temperature for each month.

INTERESTED IN PARTICIPATING IN ON-FARM RESEARCH?

Interested in working with Purdue University to address management questions and improve your operation through on-farm research? Both the Purdue Corn Agronomy Team and the Purdue On the Farm Program continue to look for on-farm cooperators for participation and assistance with on-farm research trials. In addition, we will work closely with you to answer specific questions that we and you may have specific to your own operation. Information and data collected are shared directly to each cooperator every step of the way. For more information, please reach out directly below:

Dan Quinn, PhD
Extension Corn Specialist
Purdue University
Email: djquinn@purdue.edu
Office: 765-494-5314

ABOUT THE AUTHOR

DANIEL QUINN is an assistant professor of agronomy and an extension corn specialist at Purdue University. His research and extension program focuses on improving corn production systems in the Midwest through large-scale and small-plot field trials. Quinn's key areas of study include yield physiology, agronomic intensification, precision technologies, nutrient management, and cover crops, with an emphasis on enhancing profitability, productivity, and sustainability in corn-based agriculture.

www.ingramcontent.com/pod-product-compliance
Lightning Source LLC
Chambersburg PA
CBHW041451210326
41599CB00004B/208